A Fiery Tale

By A. Skeptic

To Rocco

t a place far away, and well into the future, a crowd of children gathered around the camp fire. They had come for story-time. On this night Grandad decided to tell them about how things were long ago in the old world, when it was ruled by the Liars, an unbelievable and shocking time in our history.

'Do you want to hear it?'

'Ooh yes, please! Tell us about the time of the Great Lie, Grandad,' said Mila, and all around her the cries of agreement rang out.

'Okay,' said the old man, as he put a new log on the fire. 'Did you know, by the way, that a fire like this beside the beach was against the law at one time?'

A grasp of astonishment and disbelief rippled through the crowd.

'Why, grandad?'

'Well, the silly people thought that it would add too much warmth to the world, which was called the Environment back then, and they were very, very frightened.'

'You mean, a campfire was against the law, because it made the whole world get hotter, even though it was miles from anywhere???' 'Oh my God, the people were so stupid!'

'Ah yes,' he said sadly, shaking his head. 'It was a time when nobody could think for themselves, and they went along blindly repeating ridiculous nonsense. All sorts of wars were the result, and widespread misery and poverty. But the politicians had only themselves to blame. They carried on a fantastic lie which they carried on to make themselves rich. But I had better start at the beginning!'

'So get comfortable, and huddle together, and hold somebody's hand, because there are scary bits to this, and some terrible ways that people were outcast if they wouldn't believe what their masters said.

It besame like religion, which we spoke about last week and where people suffered just for having different ideas'

He waited a few minutes while everybody settled down, eagerly listening to every word that he said. The old man paused, as he put a big log on the fire. It lit up the night all around. In his own time and not so long ago, such a fire was rarely seen.

Taking a deep breath and speaking quietly at first, he began.

It was many, many years ago now, and once upon a time, when the kingdoms of the world were ruled by the Rulers, Princes and Dukes, like Prince Big-Ears and Duke Foot-in-mouth, who told the most outrageous lies. And even though all the poor people gave most of their money to the Rulers, these, who lived in palaces that had gold in every room, decided it was not enough and they wanted more.

Baggy Scratch'er lived in Inky Land, and she didn't much like the Minors who were dirty. 'Eww, I don't want to speak to yeeeew!' she would say, when she saw one of them in the street. But the Minors were kind and used to gather coal for her castle, although she couldn't understand why everybody couldn't have their own castle just like she did.

It so happened one time in Inky land that Baggy Scratch'er met the New Clear boys. The new clear boys promised her that they would make her castle warm and friendly and she wouldn't have to see the Minors with their black faces and hear their rough voices *ever* again.

'Oh goody', she said and she clapped her little fat hands because the Minors had a mean way of talking, and if she didn't have to see or speak to them again it would be good riddance.

The new clear boys made her a deal. If she would help them sneak into in Inky land, with their Big Secret Machines, they would make the Minor people disappear.

It seemed like a good idea to Baggy Scratch'er. And the new clear boys made up an extra big fat lie. They said they had recently discovered that all of the black coal on the fire turned the air green, and if the people kept on making coal fires, slowly all of Inky land would turn green and nothing else would be seen. Besides, people would eat the green, and turn green, and eventually would be able to see nothing but green on green everywhere.

'Well we can't have that', she said, and she lay awake at night dreaming of all the green that would lay around her castle. It gave her nightmares and a sore head thinking about, so she agreed at once to let the new clear boys do whatever they wanted.

She looked the other way while the Big Secret Machines were put in. Meanwhile they showed her all the green that was appearing in Inky land, particularly around her own castle, which made her freak out to the max and which made her hate all the people who made fires the old way. But the green had always been there, it's just that she hadn't noticed, and couldn't blame anyone, but now she blamed it on the Minors.

Instead of coal they had a special stuff called Sustainability, which same from the big secret machines, that they promised would keep her castle warm. The trouble was, nobody could see it, and yet they *imagined* that the air was turning green from the coal, just because they said so, and everybody got sucked in. But nobody could remember what it was like before. They looked around and saw Green everywhere and on everything. And they swore they'd never seen that until now.

Even if they didn't, the old green things were blamed for being newly green, and before long, all the people were marching in the street saying fires should be put out, in case the country went green. There were fights in the street between the Minors and the new clear boys, who got other people to fight for them.

Meanwhile in the Kingdom of Amer there grew to be a very powerful baron called N-Ron, who made friends with Prince Gory-Bits, and talked him meeting some New Clear boys himself. And Prince Gory-Bits (who had crowned himself prince of the Bigshots) believed in a new lie.

His lie was that the world was soon to be ruined by a nasty evil thing called GlowBall Wormy which would boil the seas, warm up the icy Poles and dry up all of life to cinders. But first of all the sea would jump out of itself.

He was especially proud of this story that he had made up to spread around, and even made a film about GlowBall Wormy and the sea rising up. Because nobody could understand it, and it was full of ridiculously wrong science, it was naturally shown to every school kid in Inkyland as fact.

Gory-Bits gave himself nightmares over his special effects every night, and got very scared about GlowBall Wormy, which he now believed himself and didn't want to stop talking about it in case it was true, so he told the other kings and queens and princes who also liked the idea of frightening people. And now even he believed it.

Some people looked around but couldn't see any seas boiling and were annoyed that they could not find this naughty GlowBall Wormy anywhere. Because Gory-Bits didn't want people to find out he might have been tricking, he sent his soldiers out to stop the unbelievers from talking. Otherwise there were certain unbelievers who rather unkindly pointed out that Gory-Bits lived in a huge house the size of a small town, which used up all the electricity.

Just then, the handsome super-hero Kyoto suddenly appeared[1] and said that if everyone gave HIM their money, he would get rid of that bad old GlowBall Wormy, which would then save the whole world from Massive Destruction.

Phew, sighed everyone, you same just in time! Lots of the Rulers like King Blah who was now in charge of Inkyland and Queen Cluck of

---

[1] In reports only, because to this day no one actually saw him

Noisyland fell instantly in love with Kyoto and wanted to give him ALL their money.

Well actually Queen Cluck was *made* to *pretend* to fall in love with Kyoto because she was desperate. The Ly-Ants party was self-destructing and she was only hanging on with one vote, so in order to stay in power she needed the Greedy party in hurry.

'Quick!' the Greedys said. 'Sign here!' It was for Kyoto Gang and because a few other countries had signed, she didn't want to be left out. She rushed to the signing table with her extra-sharpened pencil. And from that time on there was no *stronger* believer in Kyoto in all the land than Cluck. It had *nothing* to do with power.

But just then Gory-Bits was challenged to a race, and he unexpectedly lost and ran away. The winner was a rich prince called Oil George. Oil-G didn't like Gory-Bits and he also didn't like Kyoto. Oil-G wanted to figure out a way of getting most of the money, for both himself and his friends.

With Gory-Bits gone, some kingdoms were wiser, and thought it would be better to wait to see if bad old GlowBall Wormy would appear first, before they joined the new clear boys. They didn't want GlowBall Wormy to start boiling the seas and heating up the world before they were ready.

One day a secret discoverer found to his surprise that the whole idea for the Kyoto also come from the old lord N-Ron. Aha! And because N-Ron right from the beginning just wanted everybody's money, Kyoto had no intention of really killing off GlowBall Wormy!

But the kings and queens who believed in Kyoto didn't want to know this, even if it was true, because the kingdoms had already started sending the rulers the money and stuff to give to Kyoto to get rid of GlowBall Wormy. Because the rulers were allowed, under the Rules, to keep a lot of this money for themselves, they didn't want all of it to stop coming.

The kings and queens and rulers and princes had started to notice that in summertime the weather got warmer. Even though it had gotten warmer in summers in the past, they now spread the fib that any time the sun got warmer it must certainly be the handiwork of GlowBall Wormy, both now and in the future.

And 'Eek, eek!' went all the people, because as far as that goes they had short memories. That's why nobody remembers what they were eating last week or wearing last Saturday. Certainly nobody remembered when it last rained, and even if it did in the last 10 years they would quickly deny it.

Then they began to notice other little things, pointed out to them by the new clear boys. For instance, the sea was *wetter* than usual. OMG!

The ice and snow, in some places *didn't fall at all*, like Rangitoto. Second OMG!! There were places where volcanoes sprang up and nobody could explain why. Sometimes there were several corruptions happening all at once around the world, all due to GlowBall Wormy!

EEK! We're all goin' to die! Nobody could explain why anyway, because in the past things just happened, but now of course this made everybody worried about what GlowBall Wormy might do next, they began to notice some lands were drying up and some other lands were getting floods.

In fact anything that was bad got blamed on GlowBall Wormy, including when the dog ate your homework and when the handle broke on the teapot. 'Eek eek!' then went the people. And because their kingdoms were giving them more and more money to find out for sure, the Rulers travelled to Inkyland to discuss the situation.

They all decided that GlowBall Wormy must be out there somewhere, if not now then maybe appearing later, like 200 years from now. Even though GlowBall Wormy was a made up thing, like Sally the Ghost, or Peter the Perpendicular, or Robert The Really Rotten Right-angle, all of them believed their own stories and whispered them to each other, just to keep the money rolling in.

Oil-G was at the Big Meeting in Inkyland and as he was now the Ruler of the Kingdom of Amer, and because he had friends who had big sticks and who liked bashing people up, everybody listened eagerly to what he said.

As it happened, Oil-G rather liked King Blah because Blah had recently helped Oil-G go and steal a lot of oil from the Kingdoms of I've-Gone-To-Stain and Eye-Rack, which they had to do because no one was looking after the oil properly. One of his close friends, So-Dam

Insane, wanted the wells *he* had won, but that quickly bored everybody. So they got rid of him.

Oil-G told Blah that GlowBall Wormy was probably indeed here, but was still asleep and no one could tell when it would wake up. This meant that when he did wake up he would probably be EXTRA powerfully angry and do naughty things, which might need EXTRA money from the people to fix it.

Blah was happy about that and still wanted to be Oil-G's friend, because also Oil-G had by now noticed other kingdoms that also weren't taking good enough care of their oil either, like Northcore-Ear and Eye-Ran.

Now, remember, Oil-G hadn't much liked Gory-Bits because Gory-Bits once nearly beat him in that race, which was, by the way, SO unfair, because along the way Gory-Bits had accused Oil-G of cheating and that nearly made Oil-G trip himself up.

So, just because it had been Gory-Bits's special favourite, Oil-G stood up and said Kyoto was now no longer needed, and everyone could now stop giving all that money to Kyoto and instead give it all to Oil-G, especially, as Oil-G told them, he had just recently gotten a better idea, and it was only now that he reminded himself that he could tell them about it.

Apparently one of Oil-G's friends had been inventing a plant called 'Ethans' which could make cars run even better, and the good news was that GlowBall Wormy didn't much like Ethans so might go away if Oil-G was to grow it.

The good part about it for Oil-G was that with Ethans you had to put more in your car, which meant it cost more money. The farmers who grew and made Ethans were Oil-G's friends and helpers, and were about to make a huge amount of cash out of selling it, but that is probably just coincidence.

Oil-G was very persuasive and said nice things about other rulers who weren't yet in the Kyoto gang, like the rulers of Straya and Injure and Shiner, and this made them all want to join the new gang of Oil-G's who were going to grow Ethans, even though the land was already used for growing food.

Food would become much more expensive and poor people would starve, but their kingdoms would give them even more money to buy Ethans from Oil-G than to pay Kyoto. Because when it went to Kyoto, guess what, the people who said they would give it to Kyoto were actually the Rulers themselves! By hiding their own money they got more money from the others!

Well, as you can guess, the old Kyoto Gang were upset and especially the kingdom of Queen Cluck's Noisyland. She still wanted the poor people to be giving heaps of money to her royal court so she could then give it to Kyoto, which was actually *her*, and she was secretly giving it to the UNworld who were saving it for when she got tired of politics and went over to *them*.

Somebody whispered to her, so she had it on good authority that Kyoto was now angry with Noisyland for not reading the rules properly, and now Noisyland and also Queen Cluck were going to be punished for it. So she needed more money to keep Kyoto quiet. She suddenly thought of exciting new ways to tax people.

And to make matters worse, many other rulers were joining the Oil-G gang, which meant that Kyoto mightn't even have enough money to even get himself properly started.

Nobody had *still* not actually *seen* Kyoto and a lot of people thought, quite rightly, that he didn't exist. But these people were made to feel sorry for themselves, and that they were a bit loony or something. Everybody told them that they would letting down the side, and that they were being very selfish and not thinking of the others. Basically they were told to shut up.

King Blah was upset because he always thought that Oil-G was his friend, and even if Oil-G wouldn't join Blah's club then he shouldn't really go and start one of his own without telling him. Some of Blah's visitors got annoyed with Blah because he couldn't work out who his friends were, so they tried to stop the trains with bombs.

What a to-do. Everybody was fighting everybody else. N-Ron had started a big money grab and the rulers had made the people take part. Now N-Ron had his head chopped off and Gory-Bits had fallen down a hole and for a while had gone where no one could hear him anymore.

Then suddenly he jumped out, holding another film he had made while in the hole, and it was a film *he* said would change everyone's minds and one that would make everybody want him to be *king*.

A lot of people said he should have stayed in the hole, but he didn't listen, and that's the inconvenient truth. By now Oil-G had completely taken over Amer and formed a new gang but the neighborhood now wasn't big enough for both the Ethans people and Kyoto Gang.

On top of that there was a rumour that the Kyoto Gang might be dying and could even be dead. So where was all the Noisyland money going? Oh dear. Queen Cluck wasn't speaking. Meanwhile the Inkyland kingdom was becoming restless and actually Queen Cluck didn't want anyone to talk about anything. Except stadiums.

The people from Shiner wanted the coal that Queen Cluck wouldn't let Noisyland people use. The trouble was that she thought GlowBall Wormy and especially Kyoto would have been watching. So she secretly swapped Noisyland coal which was doing nothing anyway, for cheap shoes made by slaves in Shiner.[2]

At the Same time she told the UNworld that everybody follow her example and not burn coal. But where on earth was the big black cloud coming from that was now all over Shiner?? Oh, dear! All the world's black fingers pointed to Cluck!

'Sssh!' said Queen Cluck. 'Shhh!' But some people were beginning to wonder once again if GlowBall Wormy was real. Yikes! It was getting desperate, and it looked like a battle was looming. Those who believed in everything didn't want to be called silly billies by their mums and those who had given all their money to Blah and Cluck didn't want to think they had been cheated.

---

[2] When the people in Beijing made a massive all black cloud above the city, they proudly said it was from Noisyland.

Meanwhile all the ones who wanted to fight GlowBall Wormy wanted everyone to join them. By now cyclones, typhoons, floods, droughts, earthquakes anything to do with the weather was called unusual, unexpected, and scary, because nobody had time to prepare for them. Also everybody loved saying 'this is the worst in living memory,' or 'this is the wettest since records began.'

Also it turned out that suddenly nobody had ever seen a fine afternoon before, or a cool morning. They were simply told they hadn't, so they hadn't. When snow arrived at the bottom of Noisyland or the top of Inkyland they all whispered behind their hands saying 'what's this white stuff? Never seen *this* before..'

Now all their time was taken up worrying about Kyoto and GlowBall Wormy and the possibility that they would have no money left anywhere, because the money for Kyoto had mysteriously disappeared.

Queen Cluck had been giving $9million of Noisylanders' money to the UNworld for safe-keeping. She ended up going over to Amer to find it, and they gave her a job spending it! Meanwhile, someone else took over Noisyland, called Wonkey.

Wonkey looked in all the cupboards and underneath the bed once Queen Cluck had gone, looking for the money, and he found some but he didn't tell anybody. Whereas before he shouted and screamed to anyone who would listen, that she was lying, suddenly Wonkey went absolutely silent, and not a word about it was spoken. Since, he spoke about her lovingly, so she would stay well away..

Afraid it may be discovered he got a box and wrote ETS on the side, into which he put the bundles of money. ETS stands for EveryThingSafe. Although he sneaked some out from time to time[3], largely he left it alone and everyone gradually forgot about it. Except him, and the money piled up.

Now it was about this time that people start to get itchy and restless. They never had any nice things because their profit was all going out to the Rulers who were supposed to be watching GlowBall Wormy and working out what he was going to do next, and that apparently was *very* expensive.

The scientists kept the money and told to tell everybody that GlowBall Wormy was alive and well. But just to be told about it wasn't enough. The scientists also said that they had been to see Kyoto and that he was well. But there were never any photos. Very, very strange don't you think?

---

[3] To pay for electoral advertising and golf trips to Hawaii

If the seas were rising up, where could it be seen? If the world was getting warm, how come snow was falling in new places? And if droughts were coming, how come it was raining so much? When these questions were put in front of scientists, they all said 'Oh I think you mean *weather*, don't you? *We're* talking about *climate*!'

It was a cunning way out of it but it kept their wallets full which was all they wanted. But nobody could find any difference between climate and weather, and when the scientists were asked the difference they lost their temper. Big time. They threw tantrums like little children and they threw their toys out of windows onto the roofs of cars.

They didn't like to be questioned, especially about their work, which they preferred to do in secret and not tell anybody else about it. They had Special Dark Cupboards which they went into to do their experiments. Nobody else was allowed to follow them. Sometimes they just sat in there and ate each other's play-lunch, and swung their legs.

One time, when there was a big earthquake in Christchurch, they went into their cupboard and then they same out and announced that they had predicted it *before*hand. They said they hadn't wanted to scare people, which was very thoughtful, and they wanted a great lot of publicity for doing all that. 'Don't forget,' they said, '*nobody* can predict earthquakes'.

'How amazing', said all the other scientists who believed them. 'You both didn't predict them, because nobody can, and yet you predicted that you *didn't* predict them a week later, after you said you *couldn't*! You truly are very clever..bravo!!'

They got $100,000 Prime Minister's Science Media prize for doing that, which was just as well because they had been hiding in the cupboards while the earthquakes were happening, and nobody had known where they went.

It was certainly hard work out for scientists. Weather changes were happening, apparently, but always somewhere else, and always when nobody noticed. Earthquakes same and went, but always predicted by somebody unqualified. They were absolutely certain about weather coming in 100 years, but just guessing (and getting it wrong) about the weather for tomorrow.

It was a hard time also to be a child in those days. The children could see how stupid the scientists were. Even tiny little children knew that when the Sun same out and warmed everything it was not unusual, surprising, or astonishing. It was because the Sun was warm, that it made the ground warm. Duh.

Everybody was telling you that it was no longer up to you to just grow up. Now your job was to look after the planet. One time a teacher told a boy it was his job to look after the *universe*. The boy was having a problem just being a boy. Having this other job given out to him did not exactly sound fair. 'Could I get some help with the heavy lifting?' he asked quietly.

There was one little boy in particular, whose name was Daniel, who everybody got to hate very quickly, except for a small circle of friends. Daniel made himself unpopular by turning up at public meetings run by scientists and asking very simple questions that the scientists could not answer, questions like how, what, why, when, where, which one, what year, and how come you're so sure?

'Leave us alone', said the scientists who wanted to be left alone. It always upset their meeting, having to answer stupid questions from little know-all nerds in glasses. The Wormers had said there were no nonbelievers, and then went hunting for the nonbelievers who they had just said didn't exist.

When they found them they called bad names, which is how he got the name Denial, in the hope that they would go away and stay away. They didn't like the questions because, as they explained, science is *not* about asking questions and getting straight answers.

Good science is about believing in fairytales and things, and like we are running out of rain, and how much the sea will boil by, and other stuff that normal people trying to feed themselves each night can't be bothered with.

For example there was a meeting in Auckland about this and that, run by the No Idea What's Ahead organisation (NIWA). 'Have your say,' said the notice, but there was no time for questions at the end, they made sure of that.

Nor was there time during the meeting for anybody else to speak.

One time the police and the guards were there to make sure nobody asked anything. Meanwhile Al-gor had just made another stupid film. And the people at NIWA rolled their eyebrows.

There was the Big Lie, and the Big Lying scientists. The Big Lying newspapers were also in on it, and people stopped buying them after a while because they were full of fake bad news that was supposed to be happening to the weather. But when they looked out the window they could not see any sign of the catastrophe.

Baggy Scratch'er was a long time dead by now, she had started it all, but the new clear boys were still very much around, and in control. That means that GlowBall Wormy was up to his old mischief but nobody took it seriously, and they decided it was time that he changed his name.

At a big meeting inside a five star hotel at a wonderful beach resort in a foreign country, all the world leaders and their friends and relatives, and politicians with their long legged young secretaries who couldn't type properly, and whose clothes were so tight it looked like tigers fighting to get out, met to discuss it.

It was supposed to be a science conference that 75,000 scientists had to attend, and could not possibly stay at home and video link to. They had to make a carbon footprint that was colossal.

But nobody was allowed in unless their name was on the list of Believers. GlowBall Wormy still did all his evil stuff, the conference people were told, but he wanted to be known now as Clum-Chang, and the people just wanted to make him happy. Although the people were a little bit confused now, they didn't dare to speak out.

So everything that happens now was due to Clum-Chang, or was it still GlowBall Wormy, or was there *two* of them now - maybe there was two of them all along! Only the rulers knew for sure, like Al-G, and he pointed to his new film that he wanted everyone to see and that he had paid a lot of money to make, and that he said was even *truer* than the last one.

People went to see what it was all about, and got freaked out to the max by all the scary stuff. They were scared even more when they learned that the ice was made of polystyrene.'

Suddenly there was a far-off rumbling, like distant thunder, and the ground shook a little bit.

'Did you hear that?' They all gripped onto each other, their eyes shining in the moon light.

'Sounds like a big old earthquake,' said the old man smiling. And then he laughed. 'Some of you looked frightened, and you looked around with wide eyes!'

'Well, we were just a bit worried,' said a girl. 'After Christchurch, you know.., we lost our home.'

'It was not the earthquake that made you leave your house,' replied the old man. 'It was the Rulers. Earthquakes happen all the time and every day. Most of the ones you never feel. New Zealand brings 15,000 earthquakes per year, that's more than 40 every day. That's several 6's in one year and one 7 every 12 years'.

'But that's not what they told us at the time', said another girl. 'They said it was a once in a millennium event, probably made more common by marking the numbers higher..'

'Ha ha haa! the old man slapped his sides and doubled over. 'They really had you sucked in! They probably also didn't tell you that all through 2011 there were more shakes in the North Island than the South Island! How do you think we get all of nature's hills, its valleys, its mountains and its plains? How do you think we get our streams and rivers and beaches? Earthquakes! Earthquakes! Earthquakes of course! Earthquakes will *never stop happening, and we never want them to. When you hear an earthquake you must think, there's nature, it's wonderful, and it's doing its thing!'*

The rumbling faded into the distance. 'But once you hear one, you will hear several more', he said. 'Don't ever be afraid of nature. Don't ever be afraid of the extreme weather. It's got every right to be here, just like <u>you</u> have!'

They could just make out a new growl, like a wild wounded animal, and it settled down once more. He sat for a little bit in silence, as if he was listening. They all felt better when it had stopped, but in a way what he was saying gave them a new perspective, and they all felt calmer.

Some of them remembered their experience so many years ago. They remembered their mothers and fathers clinging to every little bit of information that they had received, but the Rulers had kept them in the dark. Idiots will appeared on the TV nightly, every jumped up pip-squeak reporter with the need to be important on TV[4].

---

[4] e.g. John Crumble who had no qualification in broadcasting, and used to interview himself

It was a long time before Christchurch recovered, and it was a little too late because most of the people had left. The thing the Rulers feared the most was that people would not spend money at the shops, but having no money anyway meant the shops and the business section had to eventually close..

'Now where was I up to? Oh yes, people all over the world, especially the young, started getting frustrated with the way they were getting no answers. Most of the young people had grown up knowing GlowBall Wormy, or thought they knew him, all their lives. But a worm that could boil the *seas*?

And now they didn't know what to believe in because of this whole new name, which was more important than what he supposedly did.

People's heads were filled with nonsense and they couldn't think straight. They believed for example, because that's what they were told, that all the oil was running out! If you were told something is scarce, then you pay a higher price for it. The Arabs had them all paying a ridiculous fee. The Arabs weren't silly.

Once the electric car same in, and besame popular, the price of batteries mysteriously went up sky-high, and the people were worse off than they have been before. And batteries can only be made using a relatively rare metal called lithium that has to be dug up expensively out of the ground.

So the situation didn't change, and now there were holes all over the Environment but it only changed when the Rulers were pushed out. They held on to their story of GlowBall Wormy till the bitter end. Remember, the Wormers had lots of money.

The old man felt a jabbing near his feet. He looked down. A little boy was poking him with a stick. *'Mister', said the small voice, 'did the poor people ever win?'*

'Not for a long while. Some had more money than others. The rulers of course had the most. Those who spent their lives collecting money, did no good for anyone else and ended up rich and very selfish.

Imagine a sick Bill Gates squirrel who collected all the nuts in the forest, and then leased them out to all the other squirrels. That's what the Rulers were like and the Greeds. Prince Big-Ears and Duke Foot-in-mouth were supporters. They could afford to be.

They told everybody what to think, what to buy and what to believe. They simply couldn't stand it if somebody thought differently to them. These people bought themselves fancy castles and shiny new cars, and laughed at the poor people who struggled to feed themselves. 'Get off the planet!' they said. 'There's too many of you, anyway'

Unkindness and ugliness was everywhere. It was not like today, when everybody cares for everybody else. In those days, neighbours would fight and communities would make war on the people around them.

In all the countries in those days, the Rulers and people like Oil-G and King Blah and Queen Cluck made war on their own people. Sometimes the rulers had secret people who were called tourists, and who went around with guns and created trouble. Everybody was upset and paid money for more jails and more police, and the country with the biggest army was the one with the richest ruler.

Those were the days when you had to carry a backpack full of pharmaceutical products, called the Justin Case range. If you walked outside you needed pills and if you walked inside you needed some more. If you opened your eyes or shut them, and if you spoke, and it was against the law to rest up for even 10 minutes. Pills had to be taken to be safe against the effects of Glowball Wormy, which were extreme weather all the time.

Why, Queen Cluck of Noisy land even said once that a warmer afternoon, as would be caused by GlowBall Wormy, would be *such* a danger to the planet, so *great*, that it was worse than a machine-gunning sniper on your roof. Therefore you *had* to buy the sun-creams she said, which were the ones researched by her Sun-burn scientists, which was the pharmaceutical arm of Kyoto.

Queen Cluck was Ruler of the Slapper party, who used to slap rules and regulations on everybody to impress Kyoto. Everybody was restricted to a 2 minute shower. Outside fires like this one was against the law. You could only use HER eco-light-bulbs. You even needed a permit, which you paid for, to drive out of the city, because of the extra emissions you caused. The police were everywhere, checking even what you put in your *own mouth*.

She got replaced by Little Andrew, a nice chap but more boring than a cornflake, then a young girl named little black Cinders who said watch out because that's what the world will be reduced to. The people were very frightened, because she smiled at them with huge teeth. They were more frightening than her taxes and the grinning wolf in Little Red Riding Hood.

But the people got cross and pulled her back, just as she was about to swallow Fat Bennett, the Human Pudding who ordered all the poor and hungry people to stop complaining.

Cinders threatened to kill off all the cows unless they promised never to poo any more, because Kyoto said that their poo was killing the planet. Silly Cinders believed them.

The poo used to be called compost and was considered good for the planet, but now under the New Rules, everything was changed. Queen Cluck had quietly slithered back into the country, pretending she had not been thrown out of the UN. She promised to help Cinders, which meant telling her what to do.

Noisyland had many too many rulers, and there was only a small amount of people. There was one king who thought he owned all the raindrops, because his ancestors had signed a thing called the Cheaty nearly 200 years ago.

The Greedys and the Slappers wanted everybody to ride horses again and not eat anything, because meat eaters and people producing food scraps were spoiling the pristine environment. They also didn't want the horses eating anything either, because they might poo and spoil the grass.

The Cash-in party, who only wanted people to vote for them so that they could stay in power, did nothing for anyone. They promised all the other parties that they could have whatever they wanted, and they never ever delivered.

Things got so bad, nobody trusted each other, and before too long there were thousands of parties, like the Dutch Nationals, the Hospital Lunatics, the Pure Water Youth, the Hungarian Vegetarians, and the Crazy Daisies who were all over 90.

The Let's-Have-A party, besame the pressure group that banned everything. Everybody had the right to tell someone else what to do, to put them in jail if they didn't. Books besame impossible to get, and if you did impossible to read, because somebody complained that small print was antisocial. Black and print was considered to be racist and was replaced by white print on white paper.

So, in order to give into them all, and so that the rulers could stay in power, Noisyland besame the country where *nobody* wanted to live. Nobody could afford to buy a house, there were no farms anymore, no more cars, no shops, no electricity, and no engines therefore no pollution. The Slappers could jump on anything that moved, and put a tax on it. That is finally what GlowBall Wormy wanted.

And in a funny way, the Clum-Changers got to live in their sad dream. All the rubbish piled up and the rats took over, killing all the birds and making the cities look like burnt-out ghost towns. People could not grow anything because sooner or later they would have to use compost.

It quickly spread around the world, and in every country, the desire to please GlowBall Wormy completely destroyed their brains. Meanwhile the Master Races of insects, the wild animals, the birds and the fish, all of whom were far more sensible and responsible than media run by people, got on with their lives without the fear that the world would end.

When an earthquake same, or a disaster with the weather, as very often has happened in the history of the planet, the animals just got on with things. The insects made their way to higher, open ground, the birds stopped chirping for a few days, and the rest just waited in sheltered places until it was over. Then they carried on as before.

The city was still and silent then. Not a thing moved except a few birds and dogs. A plastic bag flew on a gust of wind, settled for a moment and flew on. Some people still lived there, and they went from house to house eating food that had been left behind. They were very poor and did not have a car with which to leave.

Most of the people had driven their cars to the airport in the hope of catching a plane to an overseas city. At first they were called climate refugees. Nobody told them that there were difficulties, because cell phones were not working. There was no electricity to charge them.

When they reached the airport they discovered that they were not allowed to leave. The planes were making emergency flights only, anxious not to create a carbon footprint. Soldiers, from the government and the rulers, ordered most people into pens, which were growing bigger every day. The pens were guarded and the gates were locked, and were getting crammed with people.

It is clear what the government's idea was. The people would die in the cages at the airport. What little food they had was rationed by hospital agencies, and even under lock and key, the people accepted that it was for their own good. They were told every day that the government was working on a plan to get the aeroplanes mobile, but that was just another lie also called Sustainability. It meant anything that you and I couldn't have because it wasn't very sustainable.

One by one the people lost hope, and although there were one or two who had doubts about being imprisoned, they were a small minority and were quickly shouted down and removed. And that's how things stood for a long time. The rulers raked in the money from the poor people, by spreading a scientific story that nobody could disprove.

It was a warm winter's day in the middle of the city. A boy stood up, and put his sword on. Cam carried a sword for cutting and stabbing things, for protection. Nobody argued with you if you were armed. Cam was 11 and his two friends, Jordi and Barney, had been living rough for a year, this time under a bridge, after their parents were taken from them.

The boys had all lived in the Same street. About a year ago, they had been hunting in the woods and same home to find their houses empty, and a quickly scribbled message was left behind for them from their parents, saying how they had to move quickly because the soldiers were here.

They did not know what happened to their parents, but because they were heard laughing, and laughing was illegal as it put more CO_2 in the air, their folks were taken away in a jeep.

Cam was the tallest, although they were all in the Same year. There was food to find, but you had to look for it. Deserted McDonald stores were good, because of the bulk supplies, but there were smaller restaurants with good kitchens and shelves still full. All these places were overrun by rats and mice, and cats had to be kicked out the way to get in, and that's where Cam's sword was useful.

Sometimes there were people still living in the restaurant, and they looked at the boys in silence. The CO_2 scare was very real. You were not allowed to shout or to sing. You had to hold your breath as often as possible. Police were still everywhere, looking for joggers[5]. Cam often shared what little food he was carrying, with others that he met, and on the whole everybody gave a helping hand.

'Boys,' said Cam, 'I've been thinking about Kyoto and his men who are fighting tooth and nail at this very moment and every single day with GlowBall Wormy and Clum-Chang. Although we have been told that the worst damage is in the countryside, it can't be any worse than this - right?'

---

[5] Exercise, shouting singing etc emitted more CO_2

'What are you saying, Cam?' This sounded like another wild idea.

The other two looked up at him. Tall and lanky for his age, Cam made good decisions, in fact they were thankful to have survived in the city because of Cam showing them how.

'As I was saying, Glowball Wormy has ruined this city, and for all we know has ruined the country too. Why don't we take a trip into the country and see for ourselves, and we might even meet up with Kyoto and his gang ! We can always come back.'

Barney, who didn't say much, got very excited. There had been nothing to do in the city, because one day was the same as the next. They were forever ducking for shelter when they heard the sound of an engine, because only the police had cars and could get petrol. It did not take them long to decide. Besides, it would be an adventure.

Jordi was willing too, and they made plans right away. The next day bright and early they set off. Their plan was to follow the motorway, not right on the empty asphalt but in streets alongside, and in that way reach the edge of the city and they would be fairly free.

They knew how cold it would be because it was cold in the city at night, and it would be even worse when they were away from the houses..

A dog followed them out of the city. Cam didn't mind, in fact he loved dogs but had never had one of his own. This dog kept his distance because he was afraid of all humans, but very slowly Cam won his trust by speaking very gently to him and approaching slowly.

They also found a calf, munching grass on somebody's lawn. 'Don't go near him, Cam!' yelled George frantically. Cam was surprised at Jordi's outburst. 'It's a *farm animal*,' continued Jordi with alarm. 'Full of *poison*!'

Cam sat him down. He was shaking like a leaf. 'Easy!' Cam said gently. 'It's just the same as a dog, or a cat. It just eats grass. '

'But you don't under*stand*! It belches and farts, don't you remember what we've been *taught*?'

Yes, Cam reflected. What we've been taught. Cam was sitting in silence. He had a puzzled look on his face, as if he was trying very hard to work something out. Like everybody else he had grown up listening to the stories about GlowBall Wormy and Kyoto, and the things that they got up to, because the horrors were told over and over by his parents, so he believed in the danger to the planet.

It was the time of Queen Cluck telling them what to do and grabbing all their money, saying it was to stop the world boiling away.

But Cam had come to a decision. He wanted to see for himself. He would find the hideout of the monsters and he would find out if it really was true that Kyoto wanted save the planet. Like every other boy, he had practised with his sword, stabbing the imaginary evils behind every corner in the dark, and he had gotten really good.

But this was becoming a very awkward time for Cam. Back in the city he had begun to ask people where these monsters could be found but he wondered why nobody could tell him. Nobody had seen them. Some pointed up to the sky and some pointed under the earth. That's a bit strange, thought Cam who was by no means a fool. He had decided to find them *himself*.

He also wasn't told much about who was winning. The fight to save the world had been going since before he was born, but was nobody keeping score? GlowBall Wormy and Kyoto seemed to be in the far distance and nobody was prepared to talk about them, apart from saying again and again that they were the enemy and the saviour of civilisation.

But now he began to see that he wasn't being told the whole truth. Things didn't add up. For example people went swimming with pets who were considered clean but would not swim with animals which were considered bad for the environment. It was not getting hotter and drier, except in the summer, and the winters were just as cold as they had always been.

No one was running out of water, because obviously the world was not running out of rain, and no one was choking from CO_2 fumes, no country was getting more hurricanes than normal, and no sea appeared to be rising.

It is good to have things to think about as you're walking along, and these questions kept coming up in his mind. Cam looked across to Willie the calf and he thought, surely the calf is not evil and does not know that he is polluting the air. Willie was trusting of humans and followed Cam everywhere. There was no way this gentle animal could do anything wrong.

The more he thought about, the more confused he became. He belched and farted after eating his meals, and he knew the other boys did the same. Why should the calf's belches be any different than his? If anything, calf farts should be better than theirs, because all the animal ate was the greenest of grass and the purist of vegetables.

In short, Cam slowly began to realise that maybe he was being told, by his teachers and by school, a great big lie.

Everybody seemed in on it, all the newspapers and all rulers making announcements every day, about how they were winning in the fight against the enemies of the Environment. But why?

Just then they heard the sound of engines and they jumped into a ditch to hide. The sound grew louder as the vehicles approached, and as they peeped out they saw several buses and trucks thundering down the road.

'Hey,' said Jordi. 'We must be near the airport, because we are getting to the south of the city. We'd better be careful'.

'Of *course*,' exclaimed Cam excitedly. 'The airport is just over the hill. I remember coming here once. I wonder if we can get close enough to have a look?'

The last of the buses rumbled through, and there were no cars coming the other way. The boys dashed across the road and ran up the hill. They were still a long way away but could just make out movements in the huge area put aside for the airport development.

As the city had gotten smaller, the air port had grown much bigger, because there was so many people who were forced to move. Nobody was free anymore. They had to carry cards and passport to just buy an ice cream, and then more cards and passports just to eat it. Finally more cards, passports, and permits were needed to throw anything in the bin afterwards. Of course the Rulers, as usual, got all this money.

'I can see the cages,' said Cam, 'and I can see the people being pushed into them. I wonder if my parents are there, and if they're okay. I wonder if people are trying to break free?'

From where he stood it all seemed peaceful. It all looked like everybody was in agreement with it. That was the sad part. It would be much easier if everybody was kept prisoner against their will, then if somebody sneaked up and broke the chains that were around the huge cages, they could let the people out.

But they obviously didn't want to leave, even though they were not getting fed properly, and nobody could do any work. It was like everybody was asleep even though they were awake. They walked around as though they were zombies in a dream. A long time ago they had given up. Now they were all silent, moving around slowly in their cages and all wearing exactly the same clothes, due to a tax on buttons.

When the new clear boys visited Baggy Scratch'er so many years ago, and wanted to get coal fires eliminated, she managed to trick everybody that the coal fumes were destroying the planet. Everybody was impressed with the new idea, and gave her money to investigate.

Of course where there is money offered, there is more money available if results keep on coming. The rulers all found out that the people, once they were scared, were easy to keep scared, by paying the scientists to scare them! It was a perfect system.

Meanwhile, the truth of the matter was hidden away. Nobody was allowed to question anything and the rich rulers were getting away with it, because they had said there were too many people in the world, except, of course, them, and they needed to reduce the number.

So *that's* why everybody was at the airport, or heading there, so they could be locked up for no reason. The more Cam thought about it, the more convinced he became.

And here he was, on his way to the country to find the ones responsible, and if possible join a gang that was helping to get rid of them, and there was absolutely nothing and no one to find. Because the enemy somewhere in the countryside didn't exist. Wow. The enemy wasn't there – it was at the airport!

Cam slowly turned to Jordi and Barney, with a very serious look on his face.

'Hey guys,' he said carefully. 'We have been had, *big* time!. We have been told this humongous, great big fat, fib right from the start. There's no way now, as I see it, that there is *any* danger, except from the people who are supposed to protect us! I reckon the Rulers have made a war on the people, and they have turned the whole country into willing prisoners - and the people have gone along with it! I still can't believe it.'

'The thing I hate most of all', he said, 'is not that the Rulers have gotten away with it, although that is bad enough, no, the thing I hate is the way we have sucked *ourselves* in, the way we have been led to *believe* stupid stuff without *questioning* it.'

He lay on his back in the grass, and he was wondering what to do next. Willie the calf was munching nearby.

'So that means' said Jordi, 'the animals *aren't* spoiling the countryside, because nobody talks about the dogs, or the cats, or the birds.'

'Or the insects', said Barney. 'Simply because you can't *tax* them.'

'Exactly!' cried Cam."Try taxing a whale – good luck'

He thought again of his parents locked up in a cage at the airport, and in that moment he wanted to rescue them and all the others, but how? What to do now? Where to go? Is it possible that they were the only boys to work out what the Rulers evil plan was? How would they get the message out?

There was no one around. There was hardly anyone left on the farms so they could walk wherever they wanted. They walked down to the sea and he stood there watching for a long time.

'I *thought* so!' said Cam to the boys. Despite all the years the sea was supposed to be rising it hasn't. Not even a whisker.

It was still winter and they were shivering, because they were freezing cold and they were supposed to be as hot as a barbecue sausage, so something wasn't right. Here and there, were farm animals that had missed the vaccines that had killed the others off, and were just calmly chewing the grass that looked beautiful because it had just rained.

Willie enjoyed the company of the other cows. They pooed and chatted and jumped. Then wee-ed then they ran about and pooed some more, because that's what cows do, belching and farting in freedom.

The grass was so lush and shiny, it looked like it had just rained. But where did the rain come from? It was supposed to be drought, but this was so weird.

Cam and the boys wore themselves out, what with running and playing in the rich grass under the welcome sunshine. They laid back on the grass, panting freely (which felt a little strange) and watched the birds flying high overhead. The clouds drifted across the sky like they had always done, just like in the pictures that he remembered from the old books.

Suddenly they heard a shout.

'Great day to be out walkin'!'

It was a man wearing old clothes. In this hand there was a curved stick, of the type the old farmers used to carry for helping the stock if a cow fell down a hole.

'Are you a farmer?' Jordi was surprised. He did not expect to see any.

'Why, yes' said the man, quietly. 'There are some of us still around, but you wouldn't know it because we hide in the hills, away from the bosses and the police. If they knew we were here, they would send out a squad and take us up to the airport!'

'You know, *we* thought that all the land would be finished by now, and the environment would be wrecked, because of Glowball Wormy! I mean, what happened to the drought and the flooding and the hot temperatures? The countryside looks perfectly fine to us '

'What drought? What flooding? All *farmers* know that those come all the time, and then they go away again. It's only the people in the cities, the people who make the laws and who tell the others what to do, that haven't a clue what happens in the countryside. So where are you from, boys?'

Cam told him the story, about how grew up learning at school and from his parents, about the mighty seas first boiling away then dying, then being poisoned by acids in the rain, then the almighty earthquake that would tip all land into the sea, then a tsunami which would completely flood the country, then how the sea-level would rise out of nothing and cover the

66

cities, then how the exhaust fumes of tiny cars would completely change the sky and kill the rainforests, and would somehow get underwater and cause the death of the Great Barrier Reef, and would suddenly get underground and cause the oil to vanish, then how the north and south poles would melt and the land would be no more because it would be covered with water, and then how (at the same time!) the land would dry out and become an endless desert.

'Don't forget the one about the dengue fever and malaria and locusts,' said Jordi. 'We learned about that too in science. Because the air is so thick and suffocating, no one is supposed to breathe. Gases from old refrigerators, five times heavier than air, were supposed to float up to destroy the ozone hole, and thousands of people then have to do be on the move.'

'Yes,' chimed in Barney, 'They found that they couldn't get off the planet, so the idea was to *keep* moving. Eventually everybody would live at the airports but nobody realised that no planes would be flying and making carbon footprints, and so they wouldn't be able to stay at the airports, either. But the people haven't been allowed back. They've been put into cages at the airport and we've got to get them out again!'

He looked across to the old man who was laughing so hard that no sound was coming out of his mouth.

'I had no idea it was that bad. They...they... told you *that*?!! That - that's what would *happen?* So that's where they all went?"

Cam felt embarrassed. Willie, who was lying on the grass next to him, gave a big fart

'Yes, well said!' The farmer was muttering and playing with his stick.

'My god, and you actually believed it???'

'We-ll, us school-kids *had* to believe it. Our job was to look after the planet, what was left of it, only they didn't tell us where we would start!'

'Look around you. Do you see any catastrophe? Do you see anything to *panic* over?' Then he started laughing again. 'Come on,' he said getting up. 'Feel like a swim?'

'Ooh, *yes*! I *love* swimming! Where is the *pool*?' Cam couldn't wait.

'Who needs a pool?' And the old man began to walk off. Cam and Willie the calf looked at each other. Then they jumped up to follow him.

With the old man in the front, followed by Cam and the boys, and finally the calf and the dog, all walked through the bush that was lush with life. The boys were amazed. It was so comfortable and warm in the midday sun, so just like the picture books that Cam had to ask;

'Has the land *always* been like this?'

'For years and years, buddy,' said the man. 'I, myself, was born here. I can tell you the country has never changed, except for a few campers, but if they clean up after them then they're also good for the land.'

They rounded the corner and same to some water in a beautiful pool. It trickled in at one end over rocks, and made a deep part with overhanging branches in the middle. Then it got shallower, and towards the other end continued to go over rocks again until it made its way down the hill on the other side.

It took the boys' breath away.

'This is my secret swimming hole!'

'Wow, that's wonderful,' gushed Cam, immediately stripping off his clothes. But the dog had beaten him to it and jumped in, swimming around so that all you could see was his head.

The boys and the old man watched with amusement. Willie the calf stood at the water's edge and drank and drank. 'I know someone who was thirsty,' Cam said. Willie lifted this tail and hot yellow pee same flying out. 'Charming,' said Cam. They all laughed again.

All of a sudden Willie could stand it no longer. With a great big splash he was in the pool too, and even though his feet touched the ground he went under and up again several times, just to cool off.

Jordi was amazed. What would his teachers have said? Here was the calf and the dog, swimming with the humans. The cow had already wee-ed near the water, and who knows what the dog would have done. But the water was pure to drink, and drink they did. There was a bit of a smell but it was the smell of the countryside, which is sweet and beautiful at the same time.

They put their head below the surface and tasted the water. It was cool and clear and Cam thought it was the most wonderful water he had ever tasted. Far better than the filtered, chemically treated tap water he was used to in the city.

Cam got out first, and didn't bother to dry himself. While he was waiting for the answers in his head, he thought he would have a bit of a wander to see where the stream same from.

He began to walk around the pool, crossed the rocks in his bare feet where the water same in and looked further up the valley. And there he saw something which made him almost too frightened to move.

'Hey guys!!' he shouted. 'Come and see *this*!'

The others got gingerly out of the water, and the stood shivering for a moment, beside him.

'What's wrong mate?' Barmey asked. Cam lifted his finger and pointed. '*That*,' he said. 'It's been lying there all this time.'

'What *is* it?' The others rubbed the water out of their eyes. They looked at where Cam was pointing. Something was lying in the water, barely 100 yards from where they had been swimming. It was some kind of animal carcass, perhaps a dead cow who died of old age.

The thought occurred to them all at once. That they had been swimming and the water was pure, because between them and the smelly animal the water flowed over the rocks and must have purified itself again.

'Nothing wrong with that,' said the old man, as she same up alongside Cam. 'We drank water from our water tanks all of my life. It was not filtered. It collected rainwater off the roof and also bird poo and possum shit as well. Then there were dead rats swimming in the tank and leaves and mud. But the water was drinkable. In fact, it was all we had.' He winked. Barney was just about to say dead rats don't swim.

But this was *not* what they had been taught *either*. For it had been always said the flowing rivers and streams were polluted by farm animals, and there was *not one* river in the country that was swimmable. It was the last straw for them in terms of the Big Environmental Lie. Someone had to do something about it, and restore truth to the people, such that they could live their lives again.

That evening they made a campfire and cooked a rabbit that the dog had chased and caught that afternoon. The night was still and they felt like the whole countryside was theirs. Pretty soon the full moon started to rise, reaching right over their heads, or so it seemed, because it lit up everything around them.

There was considerable work to be done. Their tiny army, and any other farmers that were around, was only a few who had realised the Truth and survived. They never again wanted to be told lies and absurd nonsense about the end of the planet, which could not possibly be true.

So surely they weren't the only ones. There must be similar bands of people who realised the truth and what had to be done. But where were they? Cam and his friends had no idea. They had to have a plan.

They would go to the airport and try to find out what had happened to their people. It would be risky because there were only three of them, and possibly hundreds of soldiers and guards.

But first, the old man invited them back to his house. 'There it is,' he said proudly. Warm as toast! Welcome to my secret hideout!!'

The boys gasped. They had turned a corner and were looking at what they thought to be a giant plastic bag, draped over a small hill, but as they got closer they realised it was hundreds of supermarket bags, each filled with earth and piled up on each other to make walls, with gaps for doors and windows.

'Wow! I haven't seen one of those bags for ages and ages! They took them away from the supermarkets because they said they were killing the oceans and the Maui Dolphins, who were putting them over their heads, nobody had anything to put their groceries in, and consequently they stopped using the supermarkets!' said Cam.

'They cost about a 1/40 of a cent to make, and some like the warehouse sold them for 10 cents each, making a profit of 64,000%. It was another chance to be racketeering, another chance to rip off the customer!' said the man.

'Yes. One supermarket chain was so serious about the evils of plastic, that you had to unwrap every single thing that you had bought and throw the plastic covering in the bin before you left the store. As absolutely everything in the shop was in plastic cartons, plastic bottles, plastic wrap and with plastic lids, it was always a huge mess to clean up if you wanted to take anything home. People soon stopped doing that!'

'But you saved it all? Well done, but I can see why you have to keep the authorities away!'

'Of course they wouldn't give me a permit for this house. They said it was a fire hazard, it was endangering the planet, but it can't catch fire 'cos the walls are made of earth, and the hundred and one other reasons just to stop me doing it.' he said. 'But I am happy in it and I'm not hurting anybody else, so it is none of their business!!'

'Bravo!!' The boys were impressed. You could see all around from the inside because the plastic was clear. Then there were all sorts of interesting things to take note of. For instance a table and chairs made out of polystyrene which was very light to move around, and comfortable.

Then there were blankets that were big plastic bags filled with tiny plastic chips that you used for filling a space, and the plastic bags were stapled at the top. But it made a beautiful duvet. They looked around the room. It was warm and solid, after all it was virtually made of earth and the bags just kept it together.

There was plenty of light, and it was cosy and friendly. Yet you couldn't see it from the outside because it was well camouflaged as a hill, built in a hollow so it was out of the wind. The roof was just more bags filled with earth, on a simple frame.

'One day they'll discover it and pull it down,' said the man. 'I'll just wait until they've gone and put it up again, it'll take me about five minutes!' he laughed.

They slipped there the night and they were very comfortable. Cam was on a bit of raised ground but he didn't need blankets. He slept like a baby.

Fresh after sausages and eggs cooked over a wonderfully smelling fire, they slowly made their way back to the airport. Altogether it took them two days because they only travelled at night so they would not be seen. It was all so quiet.

'Shhh!!' They crept along the hills surrounding the airport and looked down on them all. Suddenly, they jumped in fright! There was *another* group in front of them with the same idea!

'Who are *you*?' Jordi said quietly.

'We were just going to ask you *the same thing*!' The boys looked at each other. They were about the same age, and together they made up a sizable group of about 20 people. As they talked they discovered that this other group had come from the town in the south, and in fact they had been walking for over a week to get here.

They, too, wanted some answers. They did not believe in Glowball Wormy but they didn't know that it was a big lie. Cam explained the whole thing and they didn't need much convincing. These were already children from a small country town who knew about country life.

'We've got to have a leader,' they said, 'and let's vote Cam!' It was a popular choice, and Cam had his sword. Cam stood up. In the half darkness he already looked like a hero who could save the world.

It was too late to turn back, and anyway what was back there? It is no life when your family are not around you. Cam knew he had to find them, and each person in the group was looking for the same thing.

Cam raised his sword high. 'Glowball Wormy is dead,' he proclaimed, 'but then it was never alive!' 'Kyoto is dead too! Hooray for common sense!!' There was big cheering around him.

They camped for the best part of the week in that spot above the airport. They could not be seen because by day they were asleep and by night they moved around and hunted for seeds. And when they were awake they could study the layout of the airport, and they saw that there were big areas where there were no police.

One night they all moved down, unseen by any of the guards. And then, bit by bit, they moved in amongst the people. Because they did it gradually nobody noticed, but they whispered about the lie, first to children, and then the children to the adults.

At first the adults were alarmed that the children had worked out some answers. But then they started to question things themselves, because they did not know what was going on in the rest of the country.

When the children were told they were also told to tell their parents of the need for secrecy, and for nobody to move until they got a signal.

Cam had a new plan. He had to get hold of the microphone that did the announcements. He waited until the airport was full of people and it was lunchtime, which meant thousands of people were in the lounge. People gathered there every day in the hope of finding planes flying.

Usually there were none and they went back to where they were camped. On this particular day Cam noticed a lady in an airline uniform, speaking into the wireless microphone. He watched her movements and noticed that she turned her head away to speak to people every now and then.

There were shops that still sold chocolate and newspapers and especially pharmacists that sold pills. Wandering in and hiding his sword under his sleeve he noticed a stack of sunglasses. Without going near them he reached out his sword and flicked a pair off the shelf, catching it neatly in his hand with nobody noticing.

Moving away unnoticed into the crowd and putting on the sunglasses he tried the same trick in the shop that sold hats. Once again, the shop was so crowded that nobody saw him. And once again he got lost in the crowd. He put on the hat.

Now with the glasses and the hat he was disguised. He moved back to where the announcer was standing, and he waited for a chance to get the microphone. He had all day and he was in no hurry. He knew she would come to the end of her shift and then he would make his move. It came sooner than he expected.

She put down the microphone just as somebody else was approaching to take her place.

'Hi,' she said 'it's pretty busy today. It will be good to put my feet up. I could do with a quick latte!' The two ladies went off to have a coffee in a nearby coffee bar and Cam seized his opportunity. He grabbed the microphone and stuffed it in his coat. Then he walked quickly to the back of the building and into an empty shed. He ducked down just in case he was seen and then he switched on the mike.

'*Ladies and gentlemen*,' he said calmly, with the voice of authority that sounded like a pilot. '*Calling all passengers,*' he continued. '*This is an important news announcement, I repeat, we have breaking news..'*

From his hiding place he looked out and could see through the slats in the wall. He could see the main passenger lounge and he could hear his own voice booming out. The entire airport all stopped what they doing and initially looked around to see who was doing the talking. There was absolute silence as they listened.

Cam's heart was pounding and he tried hard to keep the nervousness out of his voice. He continued.

'I repeat we have breaking news...a report has just come in from NASA that Glowball Wormy is finished, I'll repeat that, Glowball Wormy is finished. You may all return to your homes. Evidence has been received that it was a massive hoax, designed by the politicians, and Gory-Bits, who are now in hiding. Please make

your way slowly out of the airport. The service stations will be open for business '

People erupted in a huge roar. 'I told you so!' and 'see!' filled the air. There was a rush towards the entry doors and the chains snapped on the cages patrolled by the guards. So many people rushed the entrances that the uniformed men could not hold them back.

Special guards kept running through the lounge looking for the speaker into the microphone. Cam walked back through the hall. In the crowd's rushing for the outside there was confusion in the booths. He quietly put back the sunglasses he'd borrowed, and the hat in the hat shop. He slipped the microphone behind a counter and he was clean. He breathed a huge sigh of relief as he calmly walked out of the terminal.

He caught up with Jordi and Barney on the footpath outside.

'You did it! You did it!' they shouted. They hugged and danced in a circle. 'Now to find our families!'

'He hadn't lied. It was finished. I mean, it didn't even begin. The rulers, politicians and scientists who had for so long misled the public with false information, and who had caused so many to devote heaps of time and energy towards fixing a non-problem, went into hiding but were quickly found out.

It was a lesson to all. No longer could the science community tell lies just to receive the research funding. No longer could you make up a political party that promised to fix the environment, because nature is far too big and too strong for any human to control or fix. Anyone who claimed such powers was setting himself up to be a God.

But it was a lesson also to the ordinary people who had blindly believed the lies they had been told, even though they were ridiculously silly. It turns out that if enough people tell the same lie, then more people will believe it. But the lie has become part of our history which we will never repeat.'

The old man rested to catch his breath. It had been a long night and some of the people got up and went to bed, happy in their heart that there was nothing to worry about, and happy that nature is a wonderful, wild and exciting thing. To be part of it and to be out camping in it was a thrill for all.

They had all sat in silent awe and watched the magnificence of the Sun going down, and then the poetry of the golden moon coming up and passing overhead, passing the millions of stars twinkling softly and silently, just as in the early hours of the day they marvelled at the fantastic sight of the Sun full of promise and wonder *what the new day breaking out above the horizon would bring, and heard the early birds of the bush and their gay song, singing about the beauty of life.*

This was camping, living rough and not caring what you looked like and what you ate. Nothing had been as wonderful or as grand as these sights, and they were all for everyone every day, and they were all absolutely free.

One by one, the weary children climbed into their sleeping bags. The storyteller looked at the ground before he himself called it a day. It was quiet, he was the last one to go to bed. There were a few bits of rubbish here and there. He gathered them in seconds and put them in a plastic bag.

ISBN-13:
978-1978057579

ISBN-10:
1978057571